★ MILITARY VEHICLES ★

TANKS

MARTY GITLIN

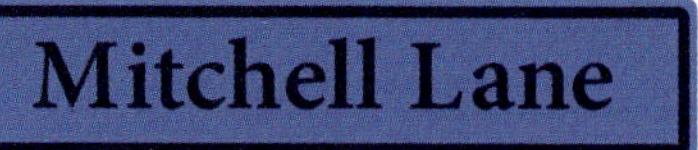

PUBLISHERS

2001 SW 31st Avenue
Hallandale, FL 33009
www.mitchelllane.com

First Edition, 2021.

Author: Marty Gitlin
Designer: Ed Morgan
Editor: Morgan Brody

Little Mitchie is an imprint of Mitchell Lane Publishers.

Title: Military Vehicles: Tanks / by Marty Gitlin
Description: Hallandale, FL :
Mitchell Lane Publishers, [2021]

Series: Military Vehicles
Library bound ISBN: 978-1-58415-730-4
eBook ISBN: 978-1-58415-733-5

Photo credits: Freepik.com, Freevector.com, Cover: U.S. Army, p. 5 U.S. Military public domain, p. 6 U.S. Navy, p. 7 Fury 1991 CC-BY-SA-4.0, p. 9 Imperial War Museum, p. 10 Imperial War Museum, p. 11 public domain, p. 13 National Archives, p. 14 U.S. Air Force, p. 17 U.S. Navy, p. 18 U.S. Air Force, p. 19 U.S. Army

CONTENTS

Words in **bold** can be found in the Glossary.

GULF WAR WIPEOUT

It was August 1990. Iraq had invaded and taken over Kuwait. Something had to be done.

The **United Nations** demanded that Iraq leave Kuwait in peace. But Iraq refused. So, a **coalition** of 39 countries banded together. They forced Iraq out of Kuwait.

American tanks **doomed** Iraq in the Battle of Medina Ridge. It began when 100 U.S. tanks rolled toward 100 Iraqi tanks.

The numbers were even. But the combat was not. United States tanks could fire almost twice as far.

Coalition forces drive an Abrams tank during the Gulf War in 1991.

American fighter jets gave support. Iraqi forces were wiped out in 40 minutes. Most of their tanks were on fire.

Iraq soon surrendered. Kuwait was saved. And it had U.S. tanks to thank for it.

A burned-out Iraqi T-55 main battle tank lies abandoned at the edge of an oil field following Operation Desert Storm.

FAST FACT

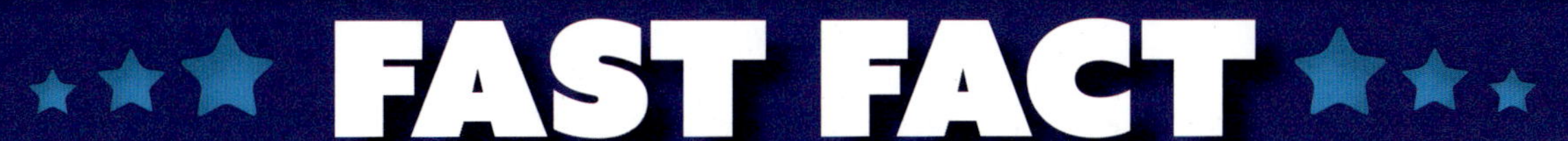

ANOTHER BIG BATTLE

Medina Ridge was not the only tank battle of the Gulf War. Perhaps the biggest was the Battle of Norfolk in February 1991. That is when the U.S. destroyed 85 Iraqi tanks. About 1,000 Iraqi troops were killed.

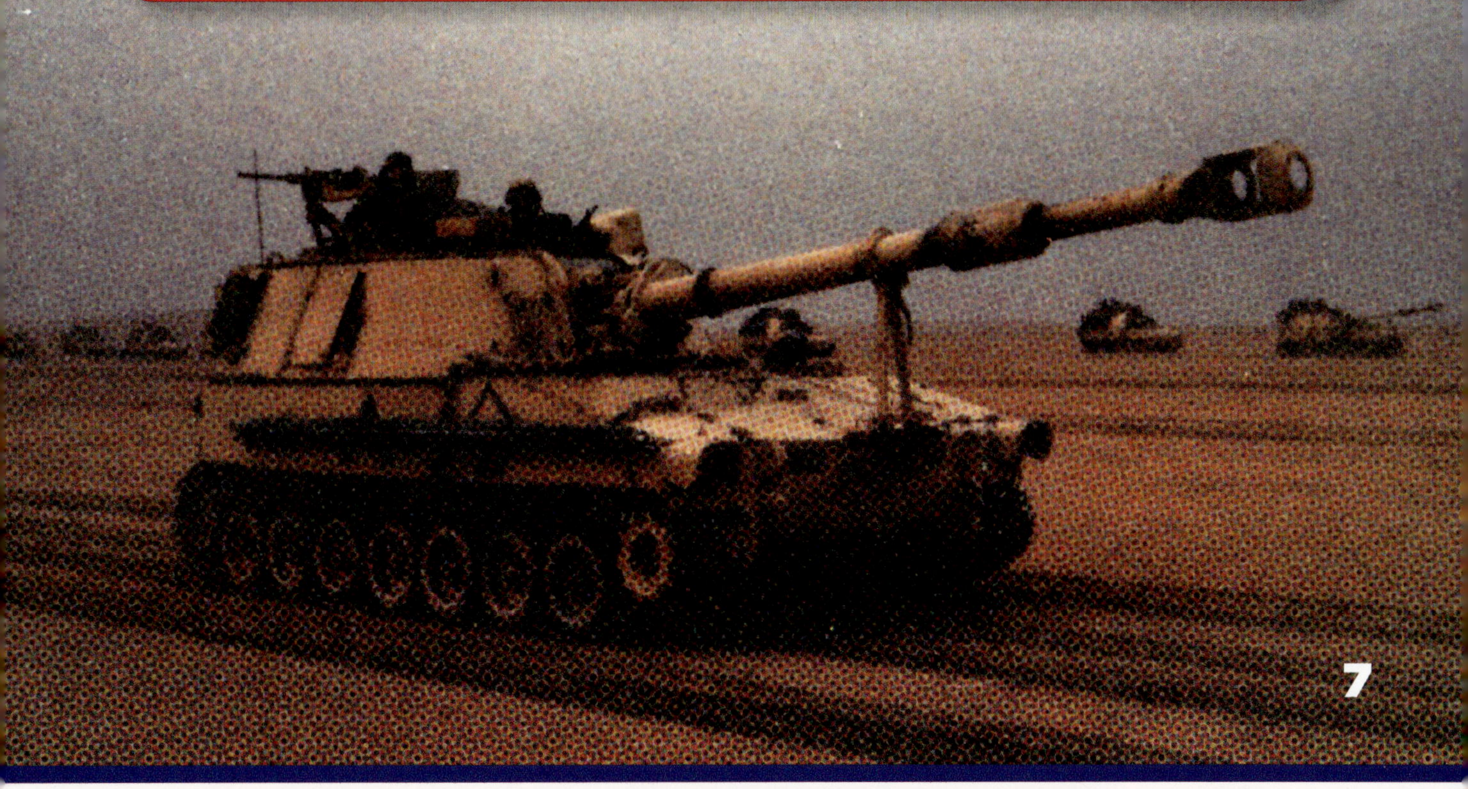

2

ROLLING INTO REALITY

It was September 1916. British troops attacked German forces in France. World War I had been raging for two years.

Suddenly new weapons rolled in. They were 32 British tanks. They had **turrets** that could not rotate. They had machine gun mounts. They were protected by **boiler plates**.

The British built tanks to win the war. Battles had always been fought on foot. It was hoped that tanks would provide firepower to destroy the enemy.

A British World War I Mark IV tank is pictured with the experimental "Tadpole Tail."

That did not happen. Only about ten tanks got through German lines. Some broke down. World War I lasted two more years.

But the British had launched a new era. Tanks changed warfare forever. They remain a force more than 100 years later.

The British F4 tank ascends a slope in 1918.

FAST FACT

A GROUP EFFORT

The U.S. helped develop a new tank at the end of World War I. The country worked with England and France to design the heavy Liberty tank. The war ended before they could be built and used.

WHY THE BEST TANKS WIN

The fastest and strongest tanks win battles. That is why American tanks have helped win wars.

Tanks fight against tanks. They roll through forests and other barriers. They have guns that can fire at targets miles away.

The U.S. Mark VIII tank takes down a tree in 1918.

The U.S. military has improved its tanks from 1916 forward. Tanks built in 1917 had rotating turrets to shoot in any direction. They went six miles an hour. And they could knock down small trees.

Tanks helped the U.S. defeat Germany in World War II. They were made with thicker **armor**. Their guns could shoot farther.

U.S. tanks won battle after battle. They rolled from France to Germany. They helped capture German cities. Soon Germany was forced to surrender.

American tanks supported the **infantry** during the Korean War. They did the same in the Vietnam War. They played bigger roles in the Gulf War and 2003 invasion of Iraq.

Soldiers plan action with their M46 Patton tank during the Korean War.

FAST FACT

BATTLE OF KURSK

The U.S. was not involved in the biggest tank battle ever. That was the 1944 Battle of Kursk in Russia. The Russians defeated Germany in that tank war. Their tanks soon rolled into Germany. Victory in World War II came shortly after.

4

THE WAY THEY WORK

Tanks don't move by themselves. They require soldiers to make them work. The U.S. Army Sherman M-1 tank needed four.

The driver controlled speed and direction. The gunner fired the two machine guns. The **ammunition** person kept the guns loaded. And the tank **commander** gave orders.

A French Army Sherman tank lands on a Normandy beach from the *USS LST-517* in August 1944.

Tanks are secured mobile vehicles. They are protected by armor. They have **platforms** on which weapons are mounted. Tanks with heavier armor are not as fast.

The U.S. M1A2 tank is among the best in the world. It fires two types of ammunition. One explodes on contact. The other uses energy to destroy enemy targets.

U.S. M1A2 Abrams tank appears ready and able.

FAST FACT

EASY ON THE DRIVER

Sherman tank drivers during World War II had many controls. Included were two steering sticks. They had a **gearbox** and a **clutch pedal**. Modern tank drivers control speed and direction with one central stick.

FUTURE OF TANK WARFARE

The ways of war keep changing. The U.S. military needs to change along with it. So must its tanks.

The military has considered fitting tanks with electric armor. That would protect them against **grenades** launched by rockets.

Other options are **stealth** technology and mobile **camouflage** systems. They could make tanks harder for enemies to find.

The military has studied many other ways to improve tanks. One is to provide them with lighter armor. That would help them **maneuver** better and go faster. Another is an energy storage system to power lasers.

It is hoped that such tanks will not be used. The goal is for their **destructive** force to stop enemies from waging war. The best way to use tanks is as weapons of peace.

FAST FACT

THE SUPER-FAST HELLCAT

The fastest tank ever might have been the Hellcat. That U.S. tank rolled at about 60 miles an hour. The World War II tank was powered by an engine like those used in aircraft. It was fast enough to race past and around slower German tanks.

What You Should Know

- Tanks have been used in war for more than 100 years.
- The British were the first to introduce tanks to warfare in 1916.
- American tanks keyed victories in World War II and the Gulf War.
- Tanks require several soldiers inside that include a driver and gunner.
- The Sherman tank could travel as fast as 60 miles an hour.
- Heavy protective armor slows tanks down.
- The Battle of Kursk during World War II is considered the greatest tank battle ever.
- Modern tanks can shoot artillery about ten miles away.
- The future could include stealth technology that makes tanks harder for radar to detect.

Glossary

ammunition
Objects fired from guns

armor
Protective covering

boiler plate
Plating of iron or steel

camouflage
To hide or disguise by covering it up or changing its looks

clutch pedal
Device used to control driving mechanisms of vehicles

coalition
A temporary union of persons, groups, or countries

commander
Person in charge who gives orders

destructive
Causing ruin

doomed
To make (someone or something) certain to fail, suffer, die, etc.

gearbox
The transmission in a vehicle

grenade
Small bomb thrown by hand or launched by rifle

infantry
Soldiers trained and equipped to fight on foot

maneuver
A clever or skillful move

platform
A level surface

stealth
Technology that allows tanks to go undetected by radar

turret
Structure on a tank in which guns are mounted

United Nations
Organization of countries dedicated to creating peace and better lives for people

Find Out More

Challen, Paul. *Mighty Tanks (Vehicles on the Move)*. Catharines, Ontario, Canada: Crabtree Publishing, 2010.

DK. *Tank: The Definitive Visual History of Armored Vehicles*. New York, N.Y.: DK Publishing, 2017.

KidCaps. *The Gulf War: A History Just For Kids!* Scotts Valley, CA: CreateSpace Independent Publishing Platform, 2013.

Websites

Ducksters: This page on the education site teaches kids about the U.S. military.
https://www.ducksters.com/history/us_government/united_states_armed_forces.php

History for Kids: Learn all about World War II tanks on this site.
https://www.historyforkids.net/world-war-2-tanks.html

Army Technology: 100 Years of Tanks. This page gives a complete history of tanks.
https://www.army-technology.com/features/featuretimeline-100-years-of-tanks-5694455/

Index

About the Author

Martin Gitlin is a veteran writer who makes his home in Cleveland, Ohio. He has authored about 150 books since 2006, mostly for young students. He has written often about tank warfare. Martin worked as a newspaper journalist from 1991 to 2002. He earned more than 45 awards during that time. He took first place for general excellence from Associated Press in 1996. That group also selected Martin as one of the top four feature writers in Ohio in 2002.